THE POETRY OF ZINC

The Poetry of Zinc

Walter the Educator™

SKB

Silent King Books a WhichHead imprint

Copyright © 2023 by Walter the Educator™

All rights reserved. No part of this book may be reproduced in any manner whatsoever without written permission except in the case of brief quotations embodied in critical articles and reviews.

First Printing, 2023

Disclaimer
This book is a literary work; poems are not about specific persons, locations, situations, and/or circumstances unless mentioned in a historical context. This book is for entertainment and informational purposes only. The author and publisher offer this information without warranties expressed or implied. No matter the grounds, neither the author nor the publisher will be accountable for any losses, injuries, or other damages caused by the reader's use of this book. The use of this book acknowledges an understanding and acceptance of this disclaimer.

"Earning a degree in chemistry changed my life!"
- Walter the Educator

dedicated to all the chemistry lovers, like myself, across the world

CONTENTS

Dedication v

Why I Created This Book? 1

One - Science And Beauty 2

Two - Vital Element 4

Three - Strength And Grace 6

Four - Mysteries Unravel 8

Five - Cherish Zinc 10

Six - Chemical Marvel 12

Seven - Purity And Grace 14

Eight - Zinc, A Treasure 16

Nine - Zinc Oxide 18

Ten - Wondrous Theme 20

Eleven - Healing And Soothing 22

Twelve - Legacy To Sustain 24

Thirteen - Enzymes, Immunity	. . . .	26
Fourteen - For In This Metal	. . . .	28
Fifteen - Whispers Of DNA	. . . .	30
Sixteen - Painter's Palette	. . . .	32
Seventeen - Beyond The Science	. . . .	34
Eighteen - Guardian Of Health	. . . .	36
Nineteen - Galvanized Armor	. . . .	38
Twenty - Silent Sentinel	. . . .	40
Twenty-One - The Protector	. . . .	42
Twenty-Two - Silvery Sheen	. . . .	44
Twenty-Three - Precious Gift	. . . .	46
Twenty-Four - Symbol Of Resilience	. . . . .	48
Twenty-Five - Tapestry Of Creation	. . . .	50
Twenty-Six - Zinc, Oh Zinc	. . . .	52
Twenty-Seven - Poems And Stories	. . . . . .	54
Twenty-Eight - Elemental Dance	. . . . . .	56
Twenty-Nine - Electric Current	. . . .	58
Thirty - Seed To Bloom	. . . .	60
Thirty-One - Batteries To Sunscreen	. . . .	62
Thirty-Two - Zinc's Brilliance	. . . .	64

Thirty-Three - Mysterious And Bright 66

Thirty-Four - Deep Fascination 68

Thirty-Five - Enzymes To DNA 70

Thirty-Six - Beauty Runs Deep 72

About The Author 74

WHY I CREATED THIS BOOK?

Creating a poetry book about the chemical element of Zinc was an interesting and unique endeavor. Zinc is a versatile and important element that plays a crucial role in various aspects of our lives, from industry and technology to health and nutrition. Exploring its properties, history, and significance through poetry can provide a fresh perspective and engage readers in a new and creative way.

ONE

SCIENCE AND BEAUTY

In the depths of Earth's embrace, Zinc does reside,
A metal of strength, with shimmering pride.
Born from the stars, in fiery creation,
This elemental wonder, a gift of elation.

Its atomic number, thirty on the chart,
Zinc, the guardian, plays a vital part.
With electrons dancing, in orbits they roam,
A symphony of nature, a celestial poem.

In mining depths, where miners toil,
Zinc is sought, beneath the soil.
A treasure hidden, waiting to be found,
A precious metal, both strong and sound.

From alloys forged, in the blacksmith's fire,
Zinc lends its strength, igniting desire.

With iron it mingles, a union so grand,
Creating a shield, against nature's demand.

In the body's tapestry, Zinc finds its place,
A guardian of health, with grace and embrace.
It weaves through cells, a protector of might,
Ensuring balance, in the realm of life's light.

From brass to batteries, Zinc's purpose is vast,
A versatile element, destined to last.
Through centuries untold, its story unfolds,
A symbol of resilience, as history beholds.

So let us celebrate, this element divine,
Zinc, a hero, with powers that shine.
In science and beauty, its legacy stands,
A testament to nature's guiding hands.

TWO

VITAL ELEMENT

Zinc, a metal so pure and bright
Shining like stars in the darkest night
Forged in the depths of the earth
A treasure of value, with countless worth
    A protector of health, a shield so strong
Defender of life, it never goes wrong
Found in nature, in rocks and soil
A giver of strength, it never spoils
    From brass to bronze, it creates alloys so fine
A conductor of electricity, it always shines
Used in galvanizing, it prevents rust
A versatile element, it's a must
    In the body, it plays a crucial role
Supporting the immune system, it's a lifesaving goal

Aiding in growth and development too
Zinc, the element, we all need it to renew

    It's fascinating how an element so small
Can have such a significant impact on us all
Zinc, a metal of wonder and might
A vital element, shining so bright.

THREE

STRENGTH AND GRACE

In the realm of metals, Zinc stands tall,
A lustrous guardian, holding a mystical call.
Its silvery hue, like moonlight's gleam,
Reflects a story, a forgotten dream.
    Beneath the surface, where secrets lie,
Zinc weaves its tale, as the days go by.
A catalyst of life, hidden in its core,
It dances with atoms, forevermore.
    From the depths of the earth, it emerges bright,
A symbol of strength, in the darkest of night.
With alchemical might, it transforms and molds,
A protector of beauty, as it unfolds.
    Zinc, the alchemist's treasure, a gift from the stars,
Unveiling its power, healing unseen scars.

In the crucible of time, it weathers and learns,
As wisdom it imparts, the world it discerns.

From the towers of temples to roofs of old,
Zinc's touch brings stories, yet untold.
A shield against rust, a guardian so true,
It stands as a sentinel, forever anew.

Oh, Zinc, the element of wonders untold,
Through ages and realms, your tale unfolds.
In your presence, we find strength and grace,
A symbol of resilience, in every space.

FOUR

MYSTERIES UNRAVEL

In the realm of chemistry, behold a metal rare,
A luminescent beauty, Zinc beyond compare.
An element of magic, it weaves a mystic spell,
A tale of transformation, a secret it will tell.

Within its silver essence, a power lies untold,
A catalyst of life, its secrets to unfold.
From sunlit skies to deepest seas, it journeys far and wide,
A silent guardian, in every place it hides.

Zinc, the alchemist's wonder, drawing spirits near,
A conductor of energy, both far and crystal clear.
In every living being, it finds its rightful place,
Nourishing the body, with elegance and grace.

From the veins of Mother Earth, it does arise,
A gift from ancient times, a treasure in disguise.

With strength and resilience, it builds a sturdy shield,
Defending flesh and bone, its presence gently healed.

In chemistry's great symphony, Zinc takes center stage,
A noble metal's dance, with elegance and grace.
Its atoms, like a ballet, moving with precision,
Creating bonds of beauty, a symphony of vision.

So, let us celebrate this element of might,
Zinc, the hidden hero, forever shining bright.
In laboratories and in nature's secret lair,
Its mysteries unravel, a tale beyond compare.

FIVE

CHERISH ZINC

In the depths of Earth's embrace, Zinc lies,
A metal with secrets, hidden from our eyes.
A guardian of balance, a true alchemist's dream,
With powers that shimmer, like a spectral gleam.

In the forge of creation, where elements collide,
Zinc binds with others, side by side.
A versatile companion, it lends a helping hand,
In the realm of chemistry, it takes a stand.

From the rusted ships sailing the sea,
To the vibrant hues of a stained-glass key,
Zinc lends its strength, a steadfast ally,
A guardian of beauty, it will never deny.

In the whispers of time, Zinc stands tall,
Through ages and eras, it conquers all.

A protector of life, within our very core,
It strengthens our bones, forevermore.

    Yet, Zinc holds mysteries, untold and pure,
A wondrous element, with tales to endure.
It weaves its magic, a silent force,
In the tapestry of life, it charts its course.

    So let us cherish Zinc, this noble treasure,
For its powers are boundless, beyond measure.
Let it remind us, in its silent grace,
That even the smallest element holds a special place.

SIX

CHEMICAL MARVEL

In a realm of metals, noble and true,
Stands Zinc, a treasure, rare and blue.
A secret force within its core,
Unveiling wonders, forevermore.

Born in the belly of Earth's deep embrace,
A silent guardian, an element of grace.
With strength concealed beneath its sheen,
Zinc, the alchemist's gleaming dream.

A shield against the corrosive might,
It guards the metals, shining bright.
An ally to iron, steadfast and strong,
Together they battle, against time's throng.

In brass it dances, a golden waltz,
Blending copper's warmth, with a touch so false.

A chameleon, a master of disguise,
Zinc, the artful imposter, weaving lies.

In batteries it pulses, a current alive,
Powering worlds, making them thrive.
A conductor of life, a conductor of might,
Zinc, the electric symphony's guiding light.

From the stars above, in cosmic delight,
Zinc descended, a celestial rite.
A gift from the heavens, a humble birth,
A testament to nature's infinite worth.

So let us marvel, at Zinc's mystic reign,
A hidden gem, beyond mundane.
A chemical marvel, a wondrous sight,
Zinc, the element, shining bright.

SEVEN

PURITY AND GRACE

In the depths of Earth's embrace, Zinc does reside,
A silent guardian, hidden in its metallic pride.
With a gleam that rivals the stars above,
It whispers tales of strength, resilience, and love.

Born from the fiery crucible of creation's core,
Zinc's essence dances with a grace to adore.
Its atomic number, thirty, a mark of distinction,
A symbol of its power, an alchemical fusion.

From ancient times, Zinc has served us well,
Unveiling secrets, its stories it does tell.
A catalyst in life's intricate chemistry,
It weaves its magic with unrivaled efficiency.

In nature's tapestry, Zinc holds its domain,
Nurturing growth, a guardian in the rain.

With a touch of Zinc, life blossoms anew,
Strengthening roots, a bond forever true.

From the depths of mines, Zinc emerges bright,
A testament to endurance, a beacon of light.
In galvanized armor, it shields us from harm,
Protecting our world with an unwavering charm.

Zinc, oh Zinc, your secrets I seek,
An enigmatic element, both bold and meek.
A symbol of balance, of purity and grace,
You illuminate our world, leaving no trace.

So, let us celebrate this element profound,
Zinc, a treasure in nature's playground.
Embrace its power, its wonders unfold,
For within its essence, miracles are told.

EIGHT

ZINC, A TREASURE

In the realm of elements, a shining star,
Zinc, the metal, known both near and far.
With atomic number thirty and a silvery hue,
It weaves its magic, revealing something new.

Beneath the surface, where mysteries reside,
Zinc's secrets unfold, like a gentle tide.
A guardian of life, it plays a vital role,
In cells and enzymes, it takes control.

From the depths of Earth, where it's found,
Zinc emerges, with purpose profound.
It lends its strength, to steel and alloy,
Standing tall, a testament to its joy.

In nature's embrace, Zinc finds its place,
In plants and creatures, it leaves its trace.

From the mightiest tree to the tiniest bee,
Zinc nurtures life, in its symphony.

But Zinc is more than just a humble aid,
Its brilliance expands, like a serenade.
In batteries and paints, it lends its spark,
A catalyst for progress, lighting the dark.

So let us celebrate this element divine,
Zinc, a treasure, in science's shrine.
For in its essence, it reminds us all,
That even the smallest can make giants fall.

In Zinc's embrace, we find inspiration,
To explore the world, with fascination.
For in the realm of elements, it's clear,
Zinc, the metal, holds magic so dear.

NINE

ZINC OXIDE

Zinc, oh mighty Zinc,
A metal that's a friend,
Used in coins and batteries,
And even for defense.

    A trace element in our body,
Necessary for good health,
Zinc keeps us strong and active,
And aids in protein synthesis.

    From the mines of earth,
Zinc is extracted with care,
It's smelted and refined,
Into a metal so rare.

    Its color is silver-blue,
With a luster that's so bright,

Zinc is an element so unique,
That's used in so many things.

It's used to galvanize iron,
And to make alloys too,
Zinc oxide is used in sunscreen,
To protect our skin from UV.

Zinc, oh mighty Zinc,
A metal that's so versatile,
From medicine to industry,
Zinc is useful in so many styles.

So let's cherish this element,
That's so important to us all,
For without Zinc, life would be different,
And our world would be quite small.

TEN

WONDROUS THEME

In a realm where metals gleam,
There lies a treasure, bright and serene.
A noble element, by nature designed,
Zinc, the alchemist's dream, refined.

    Born from the earth, its origins pure,
A silent guardian, steadfast and sure.
With atomic power, it weaves its spell,
A secret alchemy, only time can tell.

    Zinc, the protector of life's fragile frame,
A guardian angel, unknown to fame.
In cells and tissues, it finds its place,
A catalyst of health, with boundless grace.

    From rusty chains to gleaming towers,
Zinc, the master, holds the power.

A shield against corrosion's cruel embrace,
Preserving beauty, with unwavering grace.

    Zinc, the conductor, in the symphony of light,
Ignites the spark, dispelling the night.
From luminous stars to the glow of the moon,
It bathes the world in a radiant boon.

    Oh, Zinc, the mighty, with strength untold,
A silent hero, with stories yet unfold.
In every atom, a tale to be told,
Of endurance and resilience, forever bold.

    So let us raise a toast, to Zinc so rare,
A mystical element, beyond compare.
In science's realm, it reigns supreme,
Zinc, the alchemist's dream, a wondrous theme.

ELEVEN

HEALING AND SOOTHING

In a world of elements, Zinc does shine,
A metal so precious, a treasure of mine.
Its lustrous silver hue, a sight to behold,
A tale of resilience, a story untold.

Zinc, the guardian of our bodies, it's true,
In enzymes and proteins, it knows what to do.
A defender of health, it keeps us strong,
From head to toe, where it truly belongs.

From brass to bronze, its alloys are grand,
In jewelry and coins, it's in high demand.
A symbol of wealth, a mark of esteem,
Zinc, the alchemist's dream, a shimmering gleam.

In the depths of the earth, it patiently lies,
Waiting to be discovered, to dazzle our eyes.

A mineral of wonders, a secret untold,
Zinc, the hidden gem, with mysteries to unfold.

    Oh Zinc, the protector, the shield of our skin,
In sunscreens and creams, your powers begin.
With healing and soothing, you mend the wounds,
A guardian angel, a remedy that astounds.

    So let us celebrate this element divine,
Zinc, the unsung hero, that's destined to shine.
A symbol of strength, a beacon of light,
In the realm of chemistry, a guiding star bright.

TWELVE

LEGACY TO SUSTAIN

In the realm of elements, Zinc does shine,
A metal so noble, a treasure divine.
With lustrous glow and a hue so bright,
It dances with brilliance, a captivating sight.

Born from the Earth, in nature's embrace,
Zinc whispers secrets, revealing its grace.
Electrons and protons, in harmony they bind,
Creating a symphony, a melody of a kind.

Zinc, the guardian, of life's fragile thread,
In enzymes and proteins, its presence widespread.
From the depths of the ocean to the heights of the sky,
Zinc weaves its magic, never questioning why.

A healer it becomes, in wounds and in pain,
With soothing touch, it mends and sustains.

An ally to the senses, it sharpens the taste,
Unveiling flavors, a culinary embrace.
 But beyond its virtues, lies a tale untold,
Of a metal so humble, yet worth more than gold.
For Zinc, in its essence, embodies the soul,
Of strength and resilience, a story to unfold.
 So let us celebrate, this element so rare,
With gratitude and wonder, we shall declare,
Zinc, the alchemist, transforming the mundane,
Into extraordinary, a legacy to sustain.

THIRTEEN

ENZYMES, IMMUNITY

In the realm of elements, Zinc does reside,
A metal so versatile, it cannot hide.
With lustrous hue and strength untold,
Zinc, a treasure within nature's mold.

Beneath the Earth's surface, it lays concealed,
Awaiting its moment to be revealed.
Mined from the depths, a gift from the soil,
Zinc, a symbol of resilience and toil.

Its atomic number, thirty it claims,
A tale of electrons, dancing in its frames.
A conductor of electricity, a shield so rare,
Zinc, a guardian against the electric flare.

From brass to batteries, its uses are vast,
In galvanization, it's unsurpassed.

Protecting iron from rust's cruel embrace,
Zinc, the defender of metals' grace.

Yet, Zinc's worth extends beyond the mundane,
It nourishes life, a crucial refrain.
For in our bodies, it plays a key role,
Zinc, a healer that makes us whole.

Enzymes, immunity, and cell growth,
Zinc, a guardian, protecting both.
A trace element, yet vital indeed,
Zinc, the catalyst of life's every need.

So let us celebrate this element pure,
Zinc, a metal with allure.
From the Earth's embrace to our very core,
Zinc, forever, we shall adore.

FOURTEEN

FOR IN THIS METAL

In the depths of Earth's embrace, a metal gleams,
A hidden treasure, a secret it seems.
Zinc, the alchemist's dream, a wonder untold,
With powers untapped, waiting to unfold.

A humble element, unassuming and shy,
Yet its brilliance shines, catching every eye.
A guardian of health, a healer's embrace,
Zinc, the elixir, in every trace.

From ancient temples to modern halls,
Its touch whispers of forgotten calls.
In brass and alloys, it finds its place,
Adding strength and grace, with every embrace.

A guardian of rust, a protector of might,
Zinc, the defender, in battles we fight.

Shielding our vessels, our ships at sea,
Guarding against corrosion, eternally.

Zinc, the conductor, in electric streams,
Guiding the currents of our wildest dreams.
A catalyst for life, a spark in the dark,
Igniting reactions, leaving their mark.

In the tapestry of elements, Zinc weaves,
A thread of magic, ensuring life never leaves.
A silent hero, steadfast and strong,
Zinc, the unsung melody in nature's song.

So let us remember, with grateful hearts,
The wonders of Zinc, in all its parts.
For in this metal, a story is told,
Of strength, of healing, of treasures untold.

FIFTEEN

WHISPERS OF DNA

In the depths of Earth's embrace, Zinc lies,
A noble metal, hidden from prying eyes.
A lustrous beauty, shimmering and bright,
A guardian of life, in both day and night.

With steadfast strength, it stands alone,
A symbol of resilience, rarely known.
Its atoms dance, a delicate choreography,
Binding with others in perfect symmetry.

From tarnished ores, it emerges pure,
A testament to endurance, to endure.
In alchemical realms, its powers unfold,
Transforming the ordinary, a tale untold.

Zinc, the alchemist's secret, in hidden realms,
Where transmutation occurs, and magic overwhelms.

A catalyst, sparking change and innovation,
Unleashing potent potential, a grand revelation.
   In nature's embrace, it finds its place,
Nurturing life with its gentle grace.
From the green fields to the depths of the sea,
Zinc's touch brings forth vitality.
   A guardian of health, it lends a hand,
In the growth of cells, a crucial demand.
From the whispers of DNA to the beating heart,
Zinc weaves its magic, playing its part.
   Oh, Zinc, a silent hero, unsung and true,
In every aspect of life, we owe so much to you.
A symbol of strength, of transformation and might,
Zinc, a precious element, forever shining bright.

SIXTEEN

PAINTER'S PALETTE

In the depths of the Earth, where secrets lie,
A shimmering metal catches the eye,
Zinc, the element, with a lustrous sheen,
A story untold, yet to be seen.

Born in the stars, forged in cosmic fire,
Zinc emerged, a gift to inspire,
Its essence infused in nature's embrace,
A symbol of strength and boundless grace.

From ancient times, its presence known,
A treasure sought, a precious stone,
Crafted in coins, a currency of old,
Zinc's value measured, its worth untold.

In alchemy's realm, a transformative force,
Zinc's alchemical powers, a mystical course,

A catalyst, a guide to change,
Unlocking secrets, rearranging the range.
 Zinc, the protector, a guardian true,
Shielding us from harm, as warriors do,
A guardian of health, a healer's delight,
Nurturing life, bringing balance to the fight.
 In the realm of art, a masterful stroke,
Zinc's pigment vibrant, its brilliance spoke,
A painter's palette, a canvas alive,
Zinc's hues dancing, a creative dive.
 So let us celebrate this metal divine,
Zinc, the jewel in nature's design,
A shining element, full of delight,
Forever glowing, in the realm of light.

SEVENTEEN

BEYOND THE SCIENCE

In realms unseen, where alchemy resides,
A mystic metal, with secrets it hides,
Zinc, the element, in nature's grand design,
A tale untold, in every atom confined.
    From the depths of Earth, it emerges bold,
A guardian of life, a story yet untold,
Its lustrous shine, like moonlit beams,
A treasure embraced in nature's dreams.
    Beneath the stars, it dances in the night,
A celestial glow, a guiding light,
In cosmic symphony, it finds its way,
A silent guardian, come what may.
    In chemistry's realm, it weaves its spell,
A catalyst, a magician, tales to tell,

Transforming matter, with a gentle touch,
A conductor of change, it wields so much.

In every cell, it plays a vital role,
A guardian of health, a healer of the soul,
From immune defenses to DNA repair,
Zinc, the alchemist, is always there.

Yet, beyond the science, a deeper truth lies,
Zinc, a symbol, for the heart that defies,
A reminder to cherish, and to hold dear,
The elements of life, that make us persevere.

So, let us honor Zinc, in its wondrous might,
An element of wonder, shining so bright,
For in its essence, we find a profound key,
To unlock the mysteries, that set us free.

EIGHTEEN

GUARDIAN OF HEALTH

In the realm of elements, Zinc holds its sway,
A metal of wonders, in shades of gray.
Born from the earth, in mines deep and wide,
A treasure concealed, where secrets reside.

With shimmering luster, it catches the light,
Zinc dances and glows, a mystical sight.
Its strength and resilience, a gift to behold,
A protector of dreams, a warrior bold.

From ancient civilizations, Zinc was revered,
Its alchemical prowess, revered and endeared.
In brass and bronze, it lends a hand,
A binder of metals, a unifying band.

Yet Zinc's true magic lies within,
For in the body, it fights to win.

A guardian of health, a healer profound,
It strengthens the body, a remedy renowned.
　From the stars above, Zinc finds its way,
A cosmic traveler, in atoms astray.
A catalyst of life, a spark of creation,
Zinc weaves its tale, through every foundation.
　Oh, Zinc, the enigma, the element rare,
With every discovery, we stand in awe and stare.
For in your essence, mysteries untold,
A story of wonders, waiting to unfold.

NINETEEN

GALVANIZED ARMOR

In the realm of elements, let us sing,
Of a metal rare, with a silvery gleam,
Zinc, the name that adorns the periodic table,
A jewel of nature, strong and stable.

A guardian of life, it quietly resides,
In the depths of the Earth, where it hides,
Yet when summoned forth, it shines with might,
A steadfast soldier, a beacon of light.

In the body's temple, it finds its place,
A catalyst of change, with grace,
Enzymes dance, their duties fulfilled,
As Zinc orchestrates, with skills untamed.

From the rustling leaves to the ocean's spray,
Zinc whispers secrets, in its own mysterious way,

It paints vibrant hues upon the canvas of life,
A silent artist, banishing strife.

In batteries and alloys, it lends its strength,
In galvanized armor, it guards at length,
Zinc, the protector, the shield from decay,
Preserving treasures in its loyal sway.

So let us raise our voices in praise,
To Zinc, the element that never betrays,
A symbol of endurance, resilience, and more,
A metal of wonders, we simply adore.

TWENTY

SILENT SENTINEL

In the depths of Earth's embrace, a treasure lies untold,
A metal strong and steadfast, its story yet unfolds.
Element of the ancients, with secrets to impart,
Zinc, a shimmering guardian, forever in my heart.

From mines of ancient wisdom, where silence reigns supreme,
Emerges this enigma, a luminescent dream.
A guardian of the cosmos, in hidden alchemy,
Zinc weaves its mystic tapestry, a wondrous symphony.

Its touch, a gentle whisper, on skin so tender and fair,
A catalyst of healing, a shield against despair.
In potions of elixirs, where life's elixir flows,
Zinc, the humble alchemist, a cure for all life's woes.

It dances with the sunlight, in autumn's golden hue,
A guardian of the harvest, a blessing pure and true.
Through the annals of time, it weathers life's cruel storms,
Zinc, the steadfast sentinel, a guardian in all forms.

Oh Zinc, the silent sentinel, your beauty shines so bright,
A humble element, yet a beacon in the night.
In every breath we take, your essence fills the air,
Zinc, the eternal guardian, forever we shall share.

TWENTY-ONE

THE PROTECTOR

In the realm of elements, stands Zinc, bold and bright,
A metal of wonders, radiant in its light.
A catalyst of life, it weaves its magic spells,
Through the world of chemistry, where its story dwells.

Oh Zinc, the protector, defender of the weak,
A shield against corrosion, a fortress that we seek.
From rooftops to shipyards, it guards with all its might,
Preserving the structures, keeping them upright.

Beneath the earth's surface, where minerals reside,
Zinc lies in abundance, ready to be prized.
In mines deep and treacherous, miners toil away,
Extracting this treasure, day after day.

Its presence in our bodies, though small, is

profound,
Supporting our immune system, keeping us safe and sound.
Enzymes it activates, in countless biochemical ways,
Aiding in digestion, ensuring our well-being stays.

Zinc, the silent hero, in alloys it does blend,
With copper and bronze, a union that won't bend.
Adorning our jewelry, shining with grace,
Zinc adds its brilliance, a smile upon each face.

So let us raise a toast, to Zinc, strong and true,
A marvel in chemistry, with wonders to pursue.
From its atomic number, thirty and fine,
Zinc, oh Zinc, forever may you shine.

TWENTY-TWO

SILVERY SHEEN

In the depths of the earth,
Zinc lies hidden from sight,
A metal that's of great worth,
Shining in the darkness of night.

It's a trace element, they say,
Essential for life to thrive,
And in many a DNA,
It plays a role to keep us alive.

From brass to galvanized steel,
Zinc's uses are vast and wide,
And when heated to a high degree,
Its blue flames light up the sky.

Zinc's properties are unique,
A metal that's both strong and light,

And with its silvery sheen,
It's a sight that's truly bright.
    So let us honor this element,
For all the wonders it can do,
From the depths of the earth it's sent,
To help us create things anew.

TWENTY-THREE

PRECIOUS GIFT

In a realm of metals, forged by fate's decree,
There lies a hidden gem, a marvel named Zinc.
A silent guardian, it whispers in the breeze,
A tale of strength and resilience, it does not sink.
  From the depths of Earth, it emerges with grace,
A gift from nature, solid and rare.
Its lustrous gleam, a gentle embrace,
A symbol of endurance, beyond compare.
  In quiet solitude, it weaves its magic thread,
Connecting worlds, bridging the gaps.
With boundless potential, it fills the voids ahead,
A beacon of hope, where darkness perhaps crept.
  A catalyst of life, it fuels the flame,
Igniting dreams and sparking innovation.

From the lab to the fields, its power remains,
A testament to its boundless determination.
  Oh Zinc, the alchemist's prized desire,
Unveiling secrets, unlocking doors.
A guardian of health, a healer to inspire,
A treasure worth cherishing, forevermore.
  So let us honor this element, so grand and true,
For in its essence, a story unfolds.
Zinc, the silent warrior, we salute you,
A precious gift, as time unfolds.

TWENTY-FOUR

SYMBOL OF RESILIENCE

In the realm of elements, behold Zinc's might,
A metal rare, shimmering in pale light.
From Earth's crust, it emerges with grace,
A gift of nature, a gleaming embrace.

In ancient times, its secrets unknown,
But alchemists sought to make it their own.
Through fires and mixtures, they toiled with zeal,
Unraveling Zinc's essence, its appeal.

A guardian it became, protecting the steel,
From rust's cruel touch, its strength revealed.
Galvanized armor, a shield so strong,
Zinc's touch preserves, it rights every wrong.

In life's intricate dance, Zinc plays its part,
In enzymes and proteins, it ignites the heart.

Catalyst of life's reactions, it aids the way,
A silent hero, in cells it holds sway.

From roofs to batteries, Zinc finds its role,
A versatile element, with a purpose so bold.
Its brilliance shines, as it weathers the test,
Unyielding and steadfast, it outshines the rest.

Oh, Zinc, divine element, we honor your worth,
A symbol of resilience, a treasure from Earth.
In your silent presence, we find strength anew,
A reminder that even the smallest can breakthrough.

So, let us raise a toast to Zinc's grand design,
A testament to nature's brilliance, so fine.
In this ode to Zinc, its praises we sing,
A metal of wonder, a majestic offering.

TWENTY-FIVE

TAPESTRY OF CREATION

In depths unseen, where shadows roam,
Lies a secret essence, a silent home.
A metal forged from heavens' grace,
Zinc, the element, adorns this space.

    A guardian of life, a quiet knight,
Zinc, the alchemist's delight.
An alchemical dance, a symphony rare,
Beneath its veil, secrets it does share.

    From ancient mines, its story unfolds,
Through whispers of time, its legend holds.
A protector of health, a healer's boon,
Zinc, the elixir of a silver moon.

    In molecules small, a power concealed,
A gift from nature, so gently revealed.

It fuels the spark of life's grand design,
A catalyst divine, a treasure so fine.

From towering mountains to the ocean's deep,
Zinc's presence lingers, its secrets to keep.
A conductor of light, a shield from strife,
Zinc, the essence that colors our life.

In every cell, a silent guardian resides,
A whisper of Zinc, where life collides.
Oh, Zinc, the element that we adore,
Your mystique and grace forevermore.

So, let us celebrate this metal true,
Zinc, the element that shines like dew.
In the tapestry of creation, you weave,
Zinc, the hidden gem, long may you cleave.

TWENTY-SIX

ZINC, OH ZINC

Zinc, oh Zinc, a metal so bright
Shining like stars in the darkest of night
A symbol of strength, a symbol of might
A precious element, a wondrous sight

From the depths of the earth, you rise
A mineral so pure, a treasure in disguise
A gift of the universe, a blessing in disguise
A friend to humanity, a hero in our eyes

You heal our wounds, you cure our ills
A remedy so potent, a cure that thrills
You boost our immunity, you strengthen our wills
A guardian of our health, a savior that fulfills

In industry, you play a vital role
A key ingredient, a component so whole
From batteries to alloys, you make us whole
A foundation of progress, a cornerstone of success

Zinc, oh Zinc, you are more than just a metal
You are a symbol of hope, a symbol of mettle
A testament to the power of nature's vessel
A reminder of the beauty that lies within the chemical level
So let us cherish you, let us honor your worth
For you are a treasure that blesses our earth
A miracle of creation, a gift of rebirth
Zinc, oh Zinc, you are a jewel of infinite worth.

TWENTY-SEVEN

POEMS AND STORIES

In the depths of Earth, where secrets hide,
A metal emerges, strong and wide.
Zinc, the element that holds its might,
Shining with brilliance, a radiant light.

From mines it's forged, a treasure untold,
Zinc, the protector, both strong and bold.
In nature's embrace, it finds its place,
A guardian of life, with grace and embrace.

Zinc, the guardian of our fragile frame,
A part of us, in every cell's domain.
It binds DNA, a molecular dance,
Guiding life's code, with intricate chance.

From rooftops high to oceans deep,
Zinc's touch, a promise it will keep.

In galvanized armor, it shields from decay,
Preserving our world, come what may.

    Oh Zinc, you gift of the Earth's embrace,
A symbol of strength, in every trace.
With a touch of magic, you heal and restore,
An element of wonder, forevermore.

    So let us celebrate, this metal so fine,
Zinc, a treasure, in each design.
In poems and stories, let its tale be told,
Zinc, the element, forever bold.

TWENTY-EIGHT

ELEMENTAL DANCE

In the realm of elements, behold dear Zinc,
A silent force, a jewel with strength unseen.
Its atoms dance, in harmony they link,
A metal's grace, a poet's loving dream.

   Zinc, the alchemist's muse, so pure and bright,
A storyteller of secrets it does hold.
From DNA's code, it shines a light,
In Zinc Fingers, the tale of life unfolds.

   Like a pickpocket's stealth, it weaves its way,
Through cells and tissues, a conductor it plays.
In the Paris Metro's crowded sway,
Zinc whispers softly, in mysterious ways.

   Oh, how the thief, driven by desire,
Seeks gold and silver, a worldly chase.

But Zinc, the silent hero, does inspire,
A dance of life, in every human trace.

So, let us honor Zinc, this humble star,
In poems, art, and science's grand array.
For in its essence, we truly are,
A tapestry of Zinc, woven each day.

With gratitude, we celebrate, we sing,
To poets who paint Zinc's hues so bright.
Their words, like Zinc, in hearts they cling,
A symphony of voices, shining with light.

Honorable mentions, in no order fair,
For your poems, the world shall cherish and adore.
In the realm of Zinc, we find solace there,
In this elemental dance, forevermore.

TWENTY-NINE

ELECTRIC CURRENT

In the depths of Earth's embrace, Zinc lies,
A metal of strength that never defies.
A lustrous beauty, both silver and blue,
An alchemist's treasure, forever true.

Zinc, the guardian of our mortal shell,
Within our bodies, it casts a spell.
A trace element, essential, so rare,
Aiding growth, repair, and utmost care.

From the mines, it emerges, raw and bold,
A story untold, yet to unfold.
With fire and heat, it transforms and molds,
Into alloys, galvanizing the old.

Zinc, the protector against corrosion's might,
Shielding steel, embracing the light.
From rooftops to bridges, its touch is seen,
Defying rust, steadfast and serene.

In batteries, Zinc's power ignites,
A source of energy, burning bright.
From clocks that tick to cars that roar,
Zinc's electric current, forever more.

But beyond its practicality, it holds grace,
A symbol of strength, in every space.
Zinc, a reminder of our boundless potential,
To adapt, evolve, and be influential.

So let us honor this silent hero,
Zinc, the element that we all know,
For in its depths, a story lives,
Of strength, protection, and what it gives.

THIRTY

SEED TO BLOOM

In the realm of elements, Zinc takes its place,
A metal of strength, with a subtle grace.
Its aura, a shimmer, a silver gleam,
A steadfast companion, in every scheme.

From the depths of the earth, it emerges bright,
A guardian of life, both day and night.
With resilience and vigor, it does unfold,
A story of Zinc, forever untold.

In the alchemy of nature, it plays its part,
Nurturing growth, with a nurturing heart.
From seed to bloom, it ignites the flame,
Enchanting the world with its gentle claim.

In the hands of time, it does not age,
A guardian of youth, on history's page.

With healing touch, it mends the weak,
A soothing balm, for those who seek.
 Yet Zinc, in its glory, holds secrets unseen,
A mystic essence, beyond what we glean.
In the realm of science, its powers unfold,
Unlocking mysteries, yet to be told.
 So let us honor Zinc, with utmost delight,
A symbol of wonder, in our world so bright.
For in its embrace, we find solace and cheer,
A humble element, cherished and dear.

THIRTY-ONE

BATTERIES TO SUNSCREEN

In the depths of the earth,
Lies a metal of great worth,
A symbol of number 30,
Zinc is its name, so sturdy and dirty.

It's an element so vital,
In our world, it's truly pivotal,
From batteries to sunscreen,
It's uses are truly unforeseen.

It's a metal that's bluish,
And it's not at all too ghoulish,
It's not as precious as gold,
But it's worth its weight in bold.

It's a metal that's so reactive,
To acids and bases, it's proactive,

It's an element that's so unique,
In its properties, it's truly sleek.
    From the ancient times to the modern,
Zinc has been used in every pattern,
From coins to galvanized steel,
It's a metal that's truly surreal.
    So let's raise a toast to Zinc,
For its uses, we shall never blink,
It's a metal that's truly divine,
And in our world, it will forever shine.

THIRTY-TWO

ZINC'S BRILLIANCE

In the realm of elements, shines Zinc's might,
A metal so precious, a gleaming light.
Within its essence, secrets reside,
A story untold, an enigma to guide.

Born from the depths, where minerals rest,
Zinc emerges, nature's precious quest.
Silent observer, in the Earth's embrace,
It quietly reveals its captivating grace.

A guardian of health, in every way,
Zinc fortifies, keeps illnesses at bay.
In cells, it dances, a molecular dance,
Aiding enzymes, enhancing life's chance.

From brass to batteries, Zinc's talents extend,
Fueling innovation, progress to ascend.
A protector of steel, against rust's cruel sting,
It shields the structures, where memories cling.

Oh, Zinc, the alchemist's hidden gem,
A treasure that's found, yet often condemned.
For in its shadows, lies a tale untold,
Of a metal so potent, yet gentle and bold.
   So, let us celebrate this element divine,
Zinc's brilliance, forever will shine.
In the depths of science, its wonders unfurled,
A testament to nature's intricate world.

THIRTY-THREE

MYSTERIOUS AND BRIGHT

In the realm of elements, hidden and bright,
There lies a metal, a potent delight.
Zinc, the guardian, the protector of all,
A mystical force, standing proud and tall.

Its atomic number, a thirty and three,
With a lustrous allure, it captures the key.
A shield against rust, a guardian of steel,
Zinc, the alchemist's secret, it reveals.

Through ages, it weaves its magic so rare,
From ancient civilizations to modern affairs.
In brass and in bronze, its touch does impart,
A touch of elegance, a work of fine art.

In the depths of nature, its presence is found,
In mines and in ores, where treasures abound.

A symbol of courage, it strengthens the weak,
Zinc, the warrior, its powers we seek.
　From cells of the body to vision so clear,
Zinc, the healer, it banishes all fear.
Its touch brings us solace, a balm to our woes,
Zinc, the alchemist, the elixir it bestows.
　Oh Zinc, the element, mysterious and bright,
A guardian of life, a beacon of light.
In the tapestry of elements, you shine so bright,
Zinc, the metal, forever our guiding light.

THIRTY-FOUR

DEEP FASCINATION

In the depths of the Earth's embrace,
A metal, bold and full of grace.
Zinc, the element, shining bright,
A star within the darkest night.

Beneath the soil, where secrets hide,
Zinc emerges, a treasure to confide.
Its hue, a silver-gray so fine,
Reflecting light with a graceful shine.

A helper in disguise it seems,
For zinc, a savior in many schemes.
From galvanizing, to alloys strong,
Zinc's utility, forever long.

In nature's realm, it plays a part,
Within the cells, it holds a heart.

For life itself, it's crucial too,
In enzymes, where it helps us through.

 Yet zinc, a paradox it holds,
For in excess, its tale unfolds.
A poison it becomes, a silent threat,
To balance, we must never forget.

 So let us treasure this wondrous metal,
With caution and awe, in every battle.
For Zinc, a force both kind and fierce,
A symbol of strength, forever near.

 In the world of chemistry, it stands alone,
A testament to the elements, yet unknown.
Zinc, the element, a marvel of creation,
Forever admired, with deep fascination.

THIRTY-FIVE

ENZYMES TO DNA

In the depths of Earth's embrace, Zinc lies,
A metal rare, a treasure undisguised.
A guardian of life, its secrets untold,
A story of strength and beauty to behold.
    From ancient mines, its essence unfurled,
A gift from nature, changing the world.
A silent hero, it battles the rust,
Protecting the metals it loves and trusts.
    In alchemy's realm, it dances and gleams,
A catalyst, awakening dreams.
For artists and craftsmen, a muse so rare,
A medium of wonders, beyond compare.
    In chemistry's lab, it sparks and ignites,
A conductor of energy, dazzling lights.

Electrons dance in a cosmic ballet,
Zinc's atomic orchestra, leading the way.
  In biology's realm, it finds its place,
Essential to life, with grace and embrace.
From enzymes to DNA, it weaves its thread,
A building block of existence, it's said.
  From rooftops to batteries, Zinc's touch prevails,
A silent companion, as time unveils.
A testament to resilience, through trials and strife,
Zinc, the element, weaving the tapestry of life.

THIRTY-SIX

BEAUTY RUNS DEEP

In the depths of Earth, where secrets reside,
A humble metal waits, with grace and pride.
Its name is Zinc, a treasure unseen,
A lustrous element, a silvery sheen.

From ores extracted, it emerges bright,
A symbol of resilience, a guiding light.
A guardian of health, it quietly abides,
In every cell and tissue, Zinc resides.

In unity with enzymes, it takes its role,
A catalyst for life, it plays a vital role.
From taste buds to tears, it's everywhere,
Zinc, the alchemist, creating wonders rare.

With strength and endurance, it stands tall,
Protecting against rust, it never shall fall.
A shield against time, it fights the decay,
Zinc, the survivor, leading the way.

In brass and alloys, it lends its strength,
Transforming the ordinary into something of length.
From coins to cathodes, it serves us well,
Zinc, the magician, with secrets to tell.

Oh, Zinc, the unsung hero of the Earth,
A guardian of life, a substance of worth.
Let us honor your presence, forever we'll keep,
For in your shimmering essence, beauty runs deep.

ABOUT THE AUTHOR

Walter the Educator is one of the pseudonyms for Walter Anderson. Formally educated in Chemistry, Business, and Education, he is an educator, an author, a diverse entrepreneur, and he is the son of a disabled war veteran. "Walter the Educator" shares his time between educating and creating. He holds interests and owns several creative projects that entertain, enlighten, enhance, and educate, hoping to inspire and motivate you.

Follow, find new works, and stay up to date
with Walter the Educator™
at WaltertheEducator.com

www.ingramcontent.com/pod-product-compliance
Lightning Source LLC
LaVergne TN
LVHW010601070526
838199LV00063BA/5034